STEM Makerspace Projects

MAKERSPACE PROJECTS FOR BUILDING SIMPLE MACHINES

BROOKS HAYS

PowerKiDS press
New York

Published in 2021 by The Rosen Publishing Group, Inc.
29 East 21st Street, New York, NY 10010

Copyright © 2021 by The Rosen Publishing Group, Inc.

All rights reserved. No part of this book may be reproduced in any form without permission in writing from the publisher, except by a reviewer.

First Edition

Editor: Danielle Haynes
Book Design: Reann Nye
Illustrator: Benjamin Humeniuk

Photo Credits: Series art (background) ShutterStockStudio/Shutterstock.com; cover a_v_d/Shutterstock.com; p. 5 Monkey Business Images/Shutterstock.com; p. 7 kali9/E+/Getty Images; p. 9 Guenter Albers/Shutterstock.com; p. 11 Marijana Batinic/Shutterstock.com; p. 12 Dmitry_Tsvetkov/Shutterstock.com; p. 13 Jacek Chabraszewski/Shutterstock.com; p. 15 LWA/Taxi/Getty Images Plus/Getty Images; p. 17 Creaturart Images/Shutterstock.com; p. 21 DeepGreen/Shutterstock.com; p. 25 Milana Tkachenko/Shutterstock.com; p. 29 Peter Mulligan/Moment/Getty Images.

Library of Congress Cataloging-in-Publication Data

Names: Hays, Brooks Butler, author.
Title: Makerspace projects for building simple machines / Brooks Hays.
Description: New York : PowerKids Press, 2021. | Series: Stem makerspace projects | Includes bibliographical references and index. | Summary: "Simple machines are the building blocks of numerous engineering wonders. Developed centuries ago, simple machines are still helping humans redirect and magnify force today. Levers, pulleys, wheels, wings, rotors and more have helped humans span the globe, scale mountains, sail the seas, bridge mighty rivers, and take to the skies. In this volume, readers will learn how different simple machines offer different mechanical advantages, as well as how they can combine to form complex machines capable of more complicated tasks. Step-by-step instructions will help young makers construct three different simple machines-machines that can be used to solve problems and make work easier, just as they were used thousands of years ago"- Provided by publisher.
Identifiers: LCCN 2019032641 | ISBN 9781725311664 (paperback) | ISBN 9781725311688 (library bound) | ISBN 9781725311671 (6 pack)
Subjects: LCSH: Simple machines–Juvenile literature.
Classification: LCC TJ147 .H385 2021 | DDC 621.8-dc23
LC record available at https://lccn.loc.gov/2019032641

Manufactured in the United States of America

CPSIA Compliance Information: Batch #CSPK20. For Further Information contact Rosen Publishing, New York, New York at 1-800-237-9932.

Find us on

CONTENTS

WHAT IS A MAKERSPACE? 4
AN INTRODUCTION TO SIMPLE MACHINES 6
HISTORY OF SIMPLE MACHINES 8
SIMPLE MACHINES CHANGE THE WORLD 10
SIMPLE MACHINES BECOME
    COMPOUND MACHINES 12
MAKE YOUR OWN SIMPLE MACHINES 14
WINCHES AND PULLEYS 16
PROJECT 1: WINCH AND PULLEY 18
THE ARCHIMEDES' SCREW 20
PROJECT 2: ARCHIMEDES' SCREW 22
THE WHEEL AND AXLE 24
PROJECT 3: WHEEL AND AXLE 26
INVENT YOUR OWN COMPOUND MACHINE 28
SOLVE PROBLEMS AND MAKE WORK EASIER . . . 30
GLOSSARY 31
INDEX 32
WEBSITES 32

WHAT IS A MAKERSPACE?

A makerspace is both a place and a movement. It's a place where all people can gather to work together and create. Makerspaces allow students to use their hands—as well as tools, **technologies**, and an array of materials—to build, make, and create. To be human is to create, and to create is to learn and understand. Makerspaces are places where it's safe to take risks and try new things, where young learners (and all learners) can meet and socialize.

"Makerspace" also is a term for a new approach to learning that brings a do-it-yourself spirit to education. The makerspace movement provides people of all ages the freedom to craft their own adventures in learning, as well as the opportunity to solve problems with their friends and classmates.

> Makerspaces are places where children and parents, students and teachers, friends and classmates, and young and old can work together to make things, solve problems, and learn.

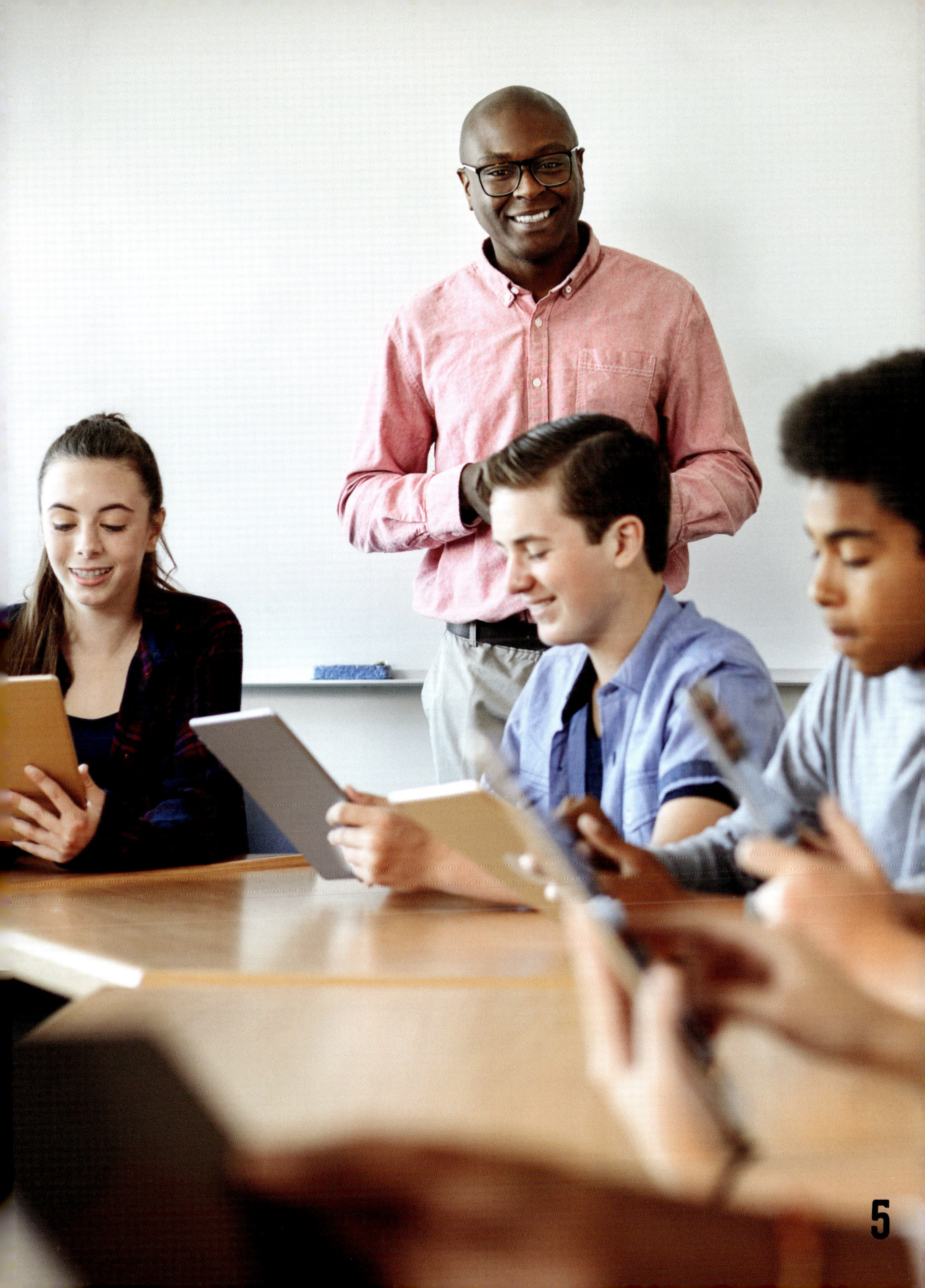

AN INTRODUCTION TO SIMPLE MACHINES

We all see and use simple machines every day. Simple machines take advantage of the physical forces that govern everything around us.

Simply put, simple machines make work easier. Like their name implies, they're not very **complex**. Simple machines in their most basic forms include pulleys, levers, wedges, screws, inclined planes, and wheels and **axles**. These simple machines are also everyday tools such as staplers, scissors, and knives, just to name a few.

Simple machines can even be used for fun. Playgrounds feature several simple machines, including seesaws and slides. Because simple machines make work easier, you can use them to solve problems. Without the use of a pulley, it would be much harder to get a flag to the top of a flagpole.

Everyday Simple Machines

You see and use simple machines every day. Levers take the form of bottle openers. Pulleys are used to raise and lower window blinds. Wedges make it easier to spread peanut butter and jelly on bread. Wheels and screws help the clock on your wall keep time. Wheelchair ramps are formed by inclined planes. You use simple machines dozens of times each day to make life easier and probably don't even realize it.

An inclined plane magnifies the amount of force you put on something, making it easier to push boxes up a ramp.

MAKER MAGIC

Mechanical advantage is the amount of help you receive through the use of a simple machine to perform work. Your dad is probably too heavy for you to pick up, but by using the help of a lever—a seesaw—you can lift him into the air!

7

HISTORY OF SIMPLE MACHINES

Humans have been using simple machines for thousands of years. Ancient people used wedges to carve stone blades—which are another kind of wedge—for hunting. The Egyptians used inclined planes to build the pyramids. Simple machines also helped early British people construct Stonehenge. Archaeologists have discovered evidence of wheels and axles in eastern Europe dating as far back as 3500 BC.

Historians credit Archimedes with first coming up with the idea of simple machines. The Greek scientist and mathematician wrote about the lever, pulley, and screw in detail. Archimedes also invented tools using simple machines, including a screw pump designed to move water vertically, from a lower point to a higher point. Today, the screw pump is known as an Archimedes' screw. Engineers still use this simple machine to move water from low places to higher places today.

MAKER MAGIC

Archimedes used his skills as an inventor to develop new weapons and technologies to defend the city of Syracuse, Greece, including a large crane-like device with a claw on the end that was used to wreck attacking ships.

Without inclined planes, wedges, levers, and other simple machines, the construction of the Egyptian pyramids wouldn't have been possible.

SIMPLE MACHINES CHANGE THE WORLD

Simple machines transformed the world long before scientists began understanding the **physics** involved. The earliest humans used levers, wedges, and inclined planes, but one of the most important simple machines in human history, the wheel and axle, didn't arrive until later, around 3500 BC.

When inventors attached the wheel and axle to a small platform, the cart was born. Carts and wagons allowed the movement of people, food, and goods across long distances. Simple machines, in other words, made trade and **efficient** long-distance travel possible.

When different groups of people trade and interact, new ideas and technologies can spread. An idea from one group can be reimagined by or combined with the idea of another, producing even more new ideas and technologies. Simple machines inspired the world's first makerspaces!

MAKER MAGIC

In a way, the first makerspaces were probably in caves, where early humans shared their ideas about how to make the best stone blades and axes.

The invention of the wheel made humans and their goods more **mobile**.

SIMPLE MACHINES BECOME COMPOUND MACHINES

Compound machines are made up of combinations of simple machines. Some compound machines, such as tractors and cars, are complex, but not all are. Many compound machines are quite simple. Think of a wheelbarrow. It combines a wheel and axle with a lever to make hauling and dumping dirt, sand, and bricks much easier. A shovel is nothing more than a wedge and lever. Still, without the basic combination of those two simple machines, clearing a snowy driveway would be a lot harder.

MAKER MAGIC

Pizza makers use a compound machine to cut slices. They combine the mechanical advantages of a wheel, wedge, and lever.

Compound machines make life easier. Try walking a mile. Then try riding your bicycle the same distance. Which one made you work harder?

If you can ride a bike, you already know how to use at least one compound machine. In addition to wheels and axles, a bike uses pulleys in the form of chains, as well as levers, which take the form of pedals.

MAKE YOUR OWN SIMPLE MACHINES

You can probably already spot simple machines all around you, but makerspaces aren't places for just looking around. They're places for action, places for making. It's in the name, after all. By making your own simple machines, you'll gain a better understanding of how simple machines make work easier. You'll also learn to identify the simple machines in the compound machines you see being used in daily life.

Makerspaces are dedicated to student-directed, hands-on learning. In the chapters ahead, you will learn about simple machines that you can make with the help of your friends and classmates. When you've completed the projects in this book, you'll have the tools and know-how to create your own one-of-a-kind compound machines by combining different simple machines.

> Making a wheel and axle using this book will make it easier for you to understand how wheels work on bigger machines, such as bicycles, cars, and airplanes.

15

WINCHES AND PULLEYS

Pulleys make lifting heavy objects easier. Pulleys consist of a flexible, or bendable, belt—often a rope or chain—wrapped around the outside of a wheel that's fixed to an axle. One end of the belt is attached to an object, while the other end is pulled to lift the object. People at your school may use a pulley to raise a flag up a flagpole.

Winches make pulling even easier. A winch is a lever attached to an axle to help turn a wheel. A rope is wrapped around the wheel, and when the winch's lever is turned, the wheel pulls on the rope.

Tow trucks use winches and pulleys to move heavy objects. On a sailboat, you can use a winch and pulley to raise a sail.

More Pulleys, Less Work For You!

The mechanical advantage offered by a pulley system can be increased by simply increasing the number of pulleys. A pulley system in which the belt is wrapped around the outside of three wheels will make it easier to lift heavy objects than a pulley system with just one wheel. Increasing the size of a wheel in a pulley system will also increase the mechanical advantage—and decrease how much force you have to use to pull!

A winch can be used to lower a bucket into and raise it out of a well.

PROJECT 1: WINCH AND PULLEY

Once you learn how to make your own winch and pulley, you'll be able to use the technology to more easily lift heavy objects.

WHAT YOU NEED

- 2 cardboard tubes of the same length
- 2 straws, one with a bend
- ribbon spool with a big enough hole in the center for a straw to pass through
- string, at least 15 inches
- tape
- scissors
- ping pong ball

WHAT YOU WILL DO

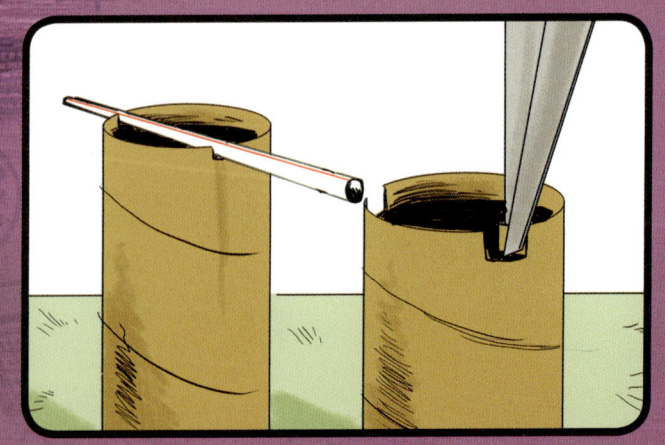

STEP 1:
On the small end of one of the cardboard tubes, use the scissors to cut two small notches on opposite sides of the opening. The notches should be wide enough for your straw to fit inside them. Repeat for second cardboard tube.

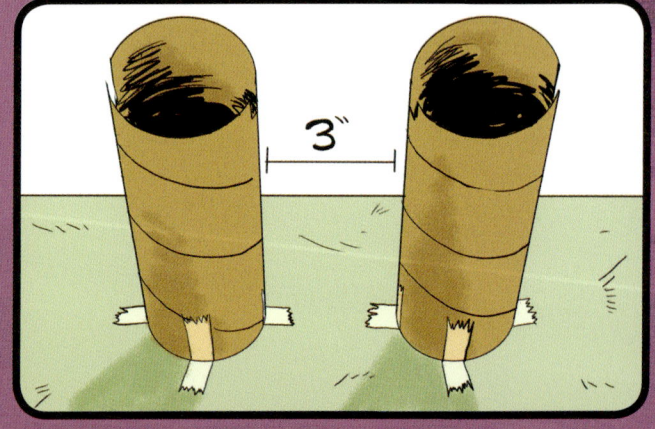

STEP 2:
With the notched ends facing up, tape the two cardboard tubes to a flat, sturdy surface like a tabletop, leaving about 3 inches of space between the tubes. Make sure the four notches are lined up.

STEP 3:
Put the straight straw through the hole in the ribbon spool. Use tape to secure the spool to the center of the straw to keep it from spinning.

STEP 4:
Rest each end of the straight straw in the notches on the cardboard tubes so the straw and ribbon spool are now supported by the tubes.

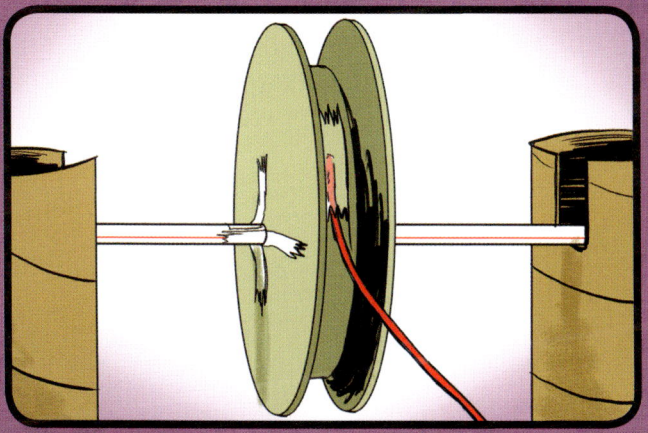

STEP 5:
Use tape to secure the string to the spool.

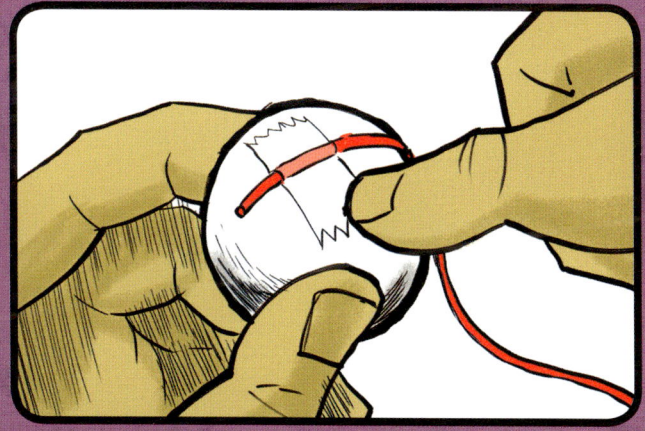

STEP 6:
Attach the other end of the string to the ping pong ball (or another small, lightweight object) using the tape.

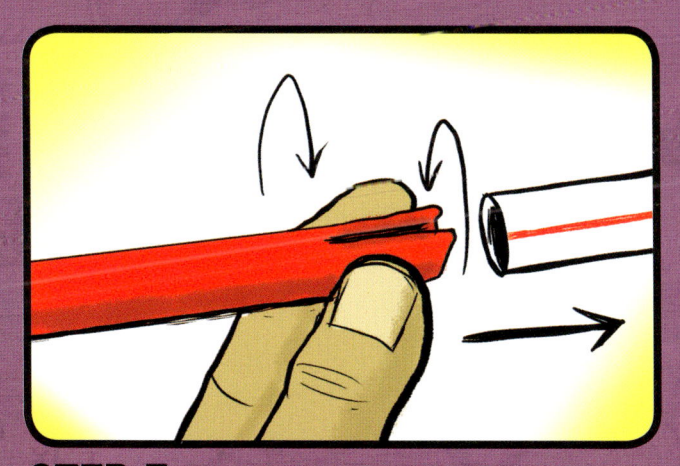

STEP 7:
Insert the long end of the bent straw into the end of the straight straw. You might have to pinch the end of the bent straw to make it fit. The small end of the bent straw now forms a handle for the winch.

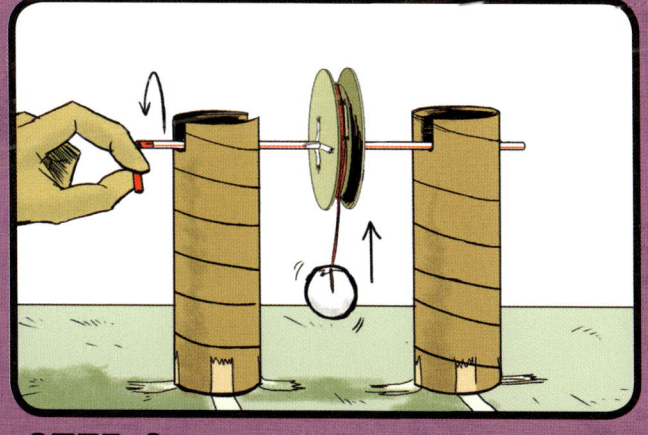

STEP 8:
Use the winch handle to turn the straw and ribbon spool, causing the string to wind around the spool and raise the ball. Turn the winch handle the other direction to lower the ball. You've now created and used a winch and pulley system!

THE ARCHIMEDES' SCREW

Greek mathematician and scientist Archimedes invented the Archimedes' screw in the third century BC. The simple machine moves water from lower to higher elevations. Early farmers used the machine to water crops.

An Archimedes' screw consists of a **helix** wrapped around a shaft and placed tightly inside a **cylinder**. The edge of the helix rests flush against the wall of the cylinder, forming sealed pockets within the cylinder. As the helix is turned, the trapped pockets of water move upward.

The Archimedes' screw, or screw pump, is still used today for irrigation or to pump water. Some historians believe Archimedes didn't invent the screw pump, but that he instead saw something like it in Egypt. These historians still credit him with popularizing the machine.

> People have been using Archimedes' screws to move water for thousands of years.

21

PROJECT 2: ARCHIMEDES' SCREW

By building and using an Archimedes' screw, you'll become a part of a technological history that stretches back to ancient times.

WHAT YOU NEED

- large mixing bowl
- cup or glass
- PVC pipe, 1.5 inch across, 14 inches long
- Flexible, clear tubing (like aquarium airline tubing), 1/4-inch inside diameter, about 4-foot length
- clear packing tape
- water
- food coloring
- scissors (optional)

WHAT YOU WILL DO

STEP 1:
Start by wrapping the plastic tubing around the outside of the PVC pipe to create a screwlike helix pattern. Leave about an inch of space between the rows of tubing.

STEP 2:
Secure the tubing to the pipe using the packing tape. Cut any excess tubing off. You now have an Archimedes' screw!

STEP 3:
Fill the large bowl with water and add food coloring of your choice.

STEP 4:
Insert the Archimedes' screw so that the bottom is submerged in the bowl of water and the other half is above the surface. The screw should be tilted at about a 45-degree angle.

STEP 5:
Place an empty glass under the top of the screw. Now, you're ready to move water.

STEP 6:
Keeping the bottom of the screw under the water, rotate the screw, making sure the opening of the clear tubing is turning toward the water. Watch the tubing scoop up some water and transport it upward as you keep turning. Water will spill out the top of the screw and into the glass as you keep turning.

THE WHEEL AND AXLE

Wheels and axles are everywhere. You're probably already aware that a wheel is a ring-shaped part secured to a rod, which is the axle. The wheel and axle rotate together. The axle is supported by a hinge or **bearing** to allow rotation.

Turning wheels reduce **friction** between an object—for example, a cart or bike—and the ground, making movement across the ground easier.

A wheel can also **amplify** a force. When you turn an axle, it increases the amount of force you use to turn a wheel. An axle can help a much bigger wheel travel greater distances. The opposite is also true—when you turn the bigger handle of a screwdriver (a wheel), it increases the force the shaft of the screwdriver (an axle) puts on a screw.

MAKER MAGIC

Around 2000 BC, humans began cutting out part of the insides of wooden wheels to create spokes, making the wheels lighter and faster.

The wheel and axle made it easier to transport heavy materials across long distances.

PROJECT 3: WHEEL AND AXLE

To demonstrate the mechanical advantage provided by a wheel and axle, we're going to build a **kinetic** sculpture. Kinetic sculptures help show motion.

WHAT YOU NEED

- three chopsticks or bamboo skewers
- cardboard box, about 7 to 9 inches wide on all sides
- wide straw cut into two 2- to 3-inch pieces
- large bottle cap or other circle-shaped item measuring 2 inches across
- pen
- scissors
- masking tape
- glue gun (to be used by an adult)

WHAT YOU WILL DO

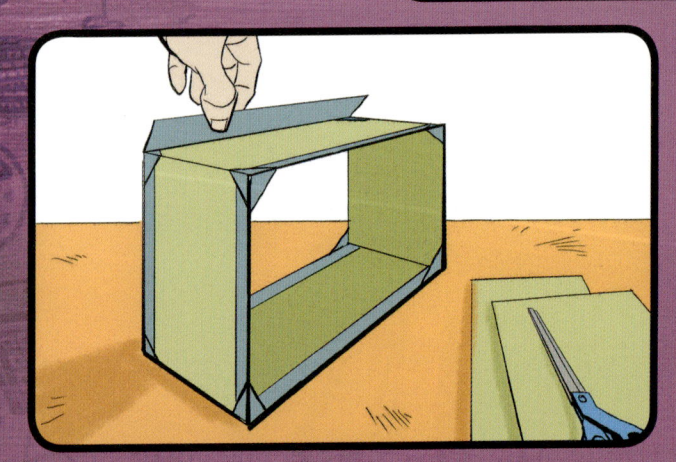

STEP 1:
Begin by cutting off the top and bottom flaps of your box to create a frame with just four sides. Save the cardboard scraps for later. Strengthen the frame with masking tape on all sides and corners.

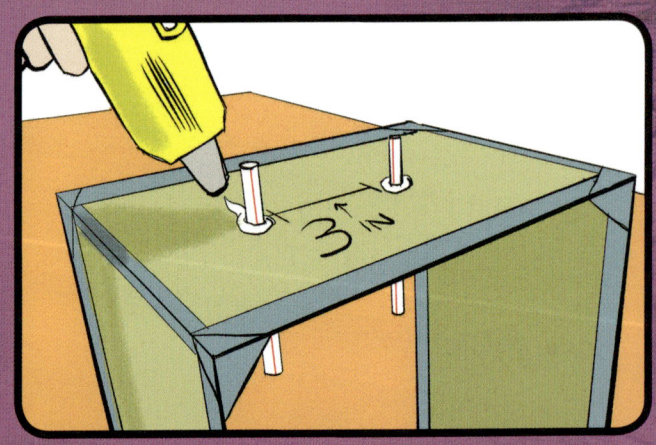

STEP 2:
On one side of your box, poke holes large enough for the straw to fit through. You may need to use a pen to widen the holes. The holes should be 3 inches apart. Stick the straws into the holes and ask an adult to help you apply hot glue around where the straws meet the box so they stay in place. This is the top side of the box.

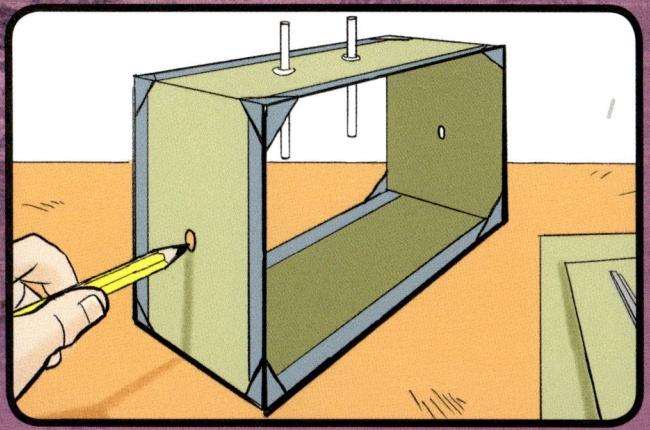

STEP 3:
On the two sides of the box, poke a hole exactly in the middle. These holes should be lined up so your chopstick/skewer can be threaded through both at the same time, going through the middle of the box.

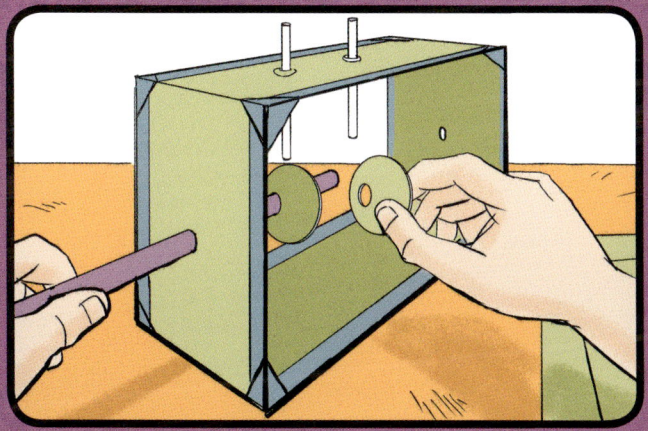

STEP 4:
Use the bottle cap to trace and cut out four circles or wheels on the leftover cardboard. Poke an off-center hole the width of a chopstick/skewer in two of the cardboard wheels.

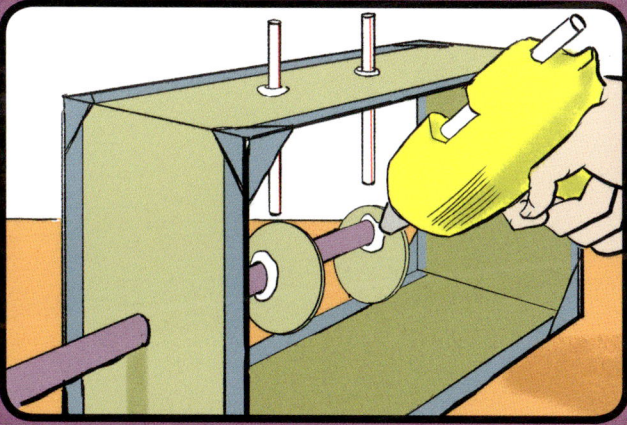

STEP 5:
Thread one chopstick/skewer through one of the holes on the box. Thread the two wheels with holes on the end of the skewer that's inside the box, then push that end of the skewer through the hole on the other side of the box.

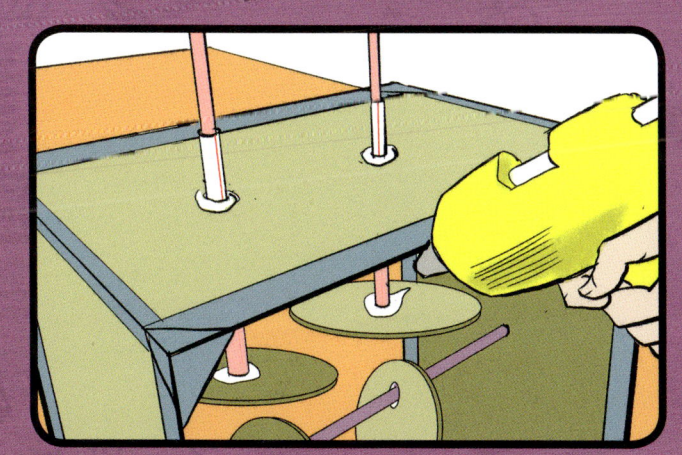

STEP 6:
Line up the two wheels so they're about 3 inches apart on the skewer, lining up with the two holes and straws in the top of the box. Ask an adult to help you apply hot glue on either side of where the wheels and the skewer meet so they don't slide.

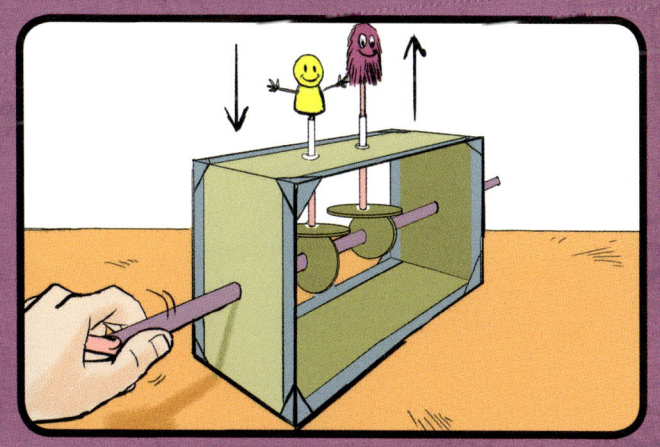

STEP 7:
Feed the two remaining chopsticks/skewers through the top two holes and straws and center the bottom ends of the skewers on the flat sides of the two remaining cardboard wheels. Apply hot glue to secure the bottoms of the skewers to the two remaining wheels.

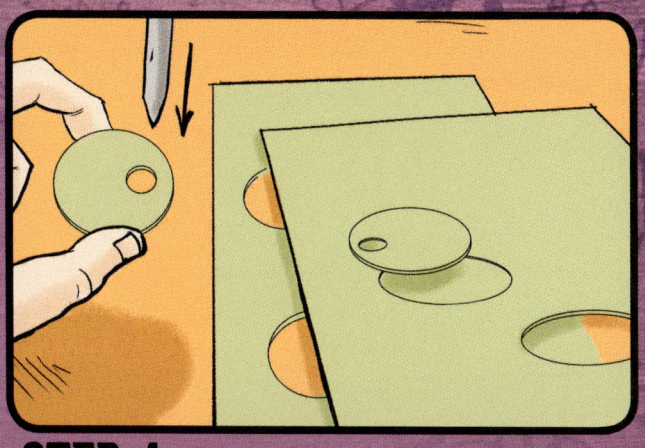

STEP 8:
Let the top two wheels rest on the rims of the bottom wheels. Now rotate the bottom skewer. As the bottom wheels turn, they should push your top skewers up and down. Try decorating the tops of the two top skewers like puppets and watch them dance up and down.

INVENT YOUR OWN COMPOUND MACHINE

You don't have to be Archimedes to develop your own compound machines. It's easy. Now that you have a working knowledge of simple machines and how to build them using everyday materials, you're ready to design and build your own compound machines.

You can combine as many simple machines as you need to solve the problem or accomplish the task you're facing, but it only takes two simple machines to create a compound machine. You can just add a lever to another simple machine to improve its efficiency. If you need your simple or compound machine to move, add some wheels and axles.

A bicycle, for instance, uses multiple simple machines—a lever for the brakes, wheels and axles, and a pulley for the chain. The possibilities are endless.

MAKER MAGIC

Though Archimedes dabbled in inventing, most of his contributions to science involved explaining the physics and mathematics of simple and compound machines. He described the how and why.

Wedges, wheels, levers, pulleys, and more can be combined to create unique compound machines like this farm tractor.

SOLVE PROBLEMS AND MAKE WORK EASIER

Work is hard. It's always been hard. But work is necessary. Humans have always had to do some basic work to survive.

Thousands of years ago, early humans used simple machines to make work easier and make survival a bit more likely. Now, you have the knowledge to join the long tradition of using simple and compound machines to make work easier.

Next time you find yourself facing a task that requires a great deal of physical effort, consider the ways a combination of simple machines might make the task easier to accomplish.

Help your parents in the garden by digging dirt with a compound machine. A shovel is a wedge and lever! Put that dirt in another compound machine. A wheelbarrow is a lever and a wheel and axle!

GLOSSARY

amplify: To increase the strength of.

axle: A bar on which wheels turn.

bearing: A part of machinery on which another part turns or moves.

complex: Having many parts.

cylinder: An object shaped like a tube.

efficient: Capable of desired results without wasting materials, time, or energy.

friction: The resistance that a surface or object encounters when moving across another.

helix: A shape formed by a line that curves around a center shape or line.

kinetic: Relating to motion.

mobile: Able to move.

physics: The science that deals with matter and energy and their interactions.

technology: A method that uses science to solve problems and the tools used to solve those problems.

INDEX

A
Archimedes, 8, 20, 22, 23, 28
Archimedes' screw, 8, 20, 22, 23
axle, 6, 8, 10, 12, 13, 14, 16, 24, 25, 26, 28, 30

C
compound machines, 12, 13, 14, 28, 29, 30

F
force, 7, 16, 24
friction, 24

I
inclined plane, 6, 7, 8, 9, 10

L
lever, 6, 7, 8, 9, 10, 12, 13, 16, 28, 29, 30

M
mechanical advantage, 7, 12, 16, 26

P
physics, 10, 28
pulley, 6, 8, 13, 16, 18, 19, 28, 29

S
screw, 6, 8, 20, 22, 23
spokes, 24

W
wedge, 6, 8, 9, 10, 12, 29, 30
wheel, 6, 8, 10, 11, 12, 14, 16, 24, 25, 26, 27, 28, 29, 30
winch, 16, 17, 18, 19

WEBSITES

Due to the changing nature of Internet links, PowerKids Press has developed an online list of websites related to the subject of this book. This site is updated regularly. Please use this link to access the list: www.powerkidslinks.com/stemmake/machines